CAREER EXPLORATION

Animal Caretaker

by Tracey Boraas

Consultant:
Nancy Peterson
Issues Specialist, Companion Animals Section
Cynthia Stitely
Outreach Coordinator, Animal Sheltering Issues
The Humane Society of the United States

CAPSTONE BOOKS
an imprint of Capstone Press
Mankato, Minnesota

Capstone Books are published by Capstone Press
151 Good Counsel Drive, P.O. Box 669, Mankato, Minnesota 56002-0669
http://www.capstone-press.com

Printed in the United States of America.

Library of Congress Cataloging-in-Publication Data
Boraas, Tracey.
Animal caretaker/by Tracey Boraas.
p. cm.—(Career exploration)
Includes bibliographical references (p. 45) and index.
Summary: Introduces the career of animal caretaker, discussing educational requirements, duties, work environment, salary, employment outlook, and possible future positions.
ISBN 0-7368-0590-7
1. Animal specialists—Vocational guidance—Juvenile literature. [1. Animal specialists—Vocational guidance. 2. Vocational guidance.] I. Title. II. Series

SF80 .B67 2001
636'.083'023—dc21 00-024692

Editorial Credits
Connie R. Colwell, editor; Steve Christensen, cover designer; Kia Bielke, production designer and illustrator; Heidi Schoof and Kimberly Danger, photo researchers

Photo Credits
Frances M. Roberts, 9
Ginger S. Buck, 6, 10, 20
Index Stock, cover, 24, 34, 40; Index Stock/Ellen Skye, 36
Leslie O'Shaughnessy, 23
Norvia Behling, 14, 26
Photri Microstock/Kulik, 38
Pictor, 32
Unicorn/David P. Dill, 12; Jim Shippee, 16; Deneve Feigh Bunde, 19; Paula J. Harrington, 28; Paul Murphy, 43

Table of Contents

Chapter 1 Animal Caretaker 7
Chapter 2 Day-to Day Activities 15
Chapter 3 The Right Candidate 27
Chapter 4 Preparing for the Career 33
Chapter 5 The Market 39

Fast Facts 4
Skills Chart.. 30
Words to Know.. 44
To Learn More.. 45
Useful Addresses 46
Internet Sites.. 47
Index 48

Fast Facts

Career Title	Animal Caretaker
O*NET Number	79017A
DOT Cluster (Dictionary of Occupational Titles)	Agricultural, fishery, forestry, and related occupations
DOT Number	410.674-010
GOE Number (Guide for Occupational Exploration)	03.03.02
NOC Number (National Occupational Classification-Canada)	648
Salary Range (U.S. Bureau of Labor Statistics and Human Resources Development Canada, late 1990s figures)	U.S.: $8,840 to $25,000 Canada: $12,480 to $35,360 (Canadian dollars)
Minimum Educational Requirements	U.S.: none Canada: none
Certification/Licensing Requirements	U.S.: none Canada: none

Subject Knowledge	Mathematics; life sciences; zoology; biology; communication skills; chemistry
Personal Abilities/Skills	Understand habits and physical needs of animals; keep calm during emergencies; observe animals to notice indicators of illness; stay patient while repeating training routines; use hands and fingers skillfully; do hard physical work
Job Outlook	U.S.: faster than average growth Canada: fair
Personal Interests	Plants and animals: interest in activities involving plants and animals, usually in an outdoor setting
Similar Types of Jobs	Agricultural and biological scientist; veterinarian; pet store worker; gamekeeper; game-farm helper; poultry breeder; rancher; artificial-breeding technician; veterinary technician

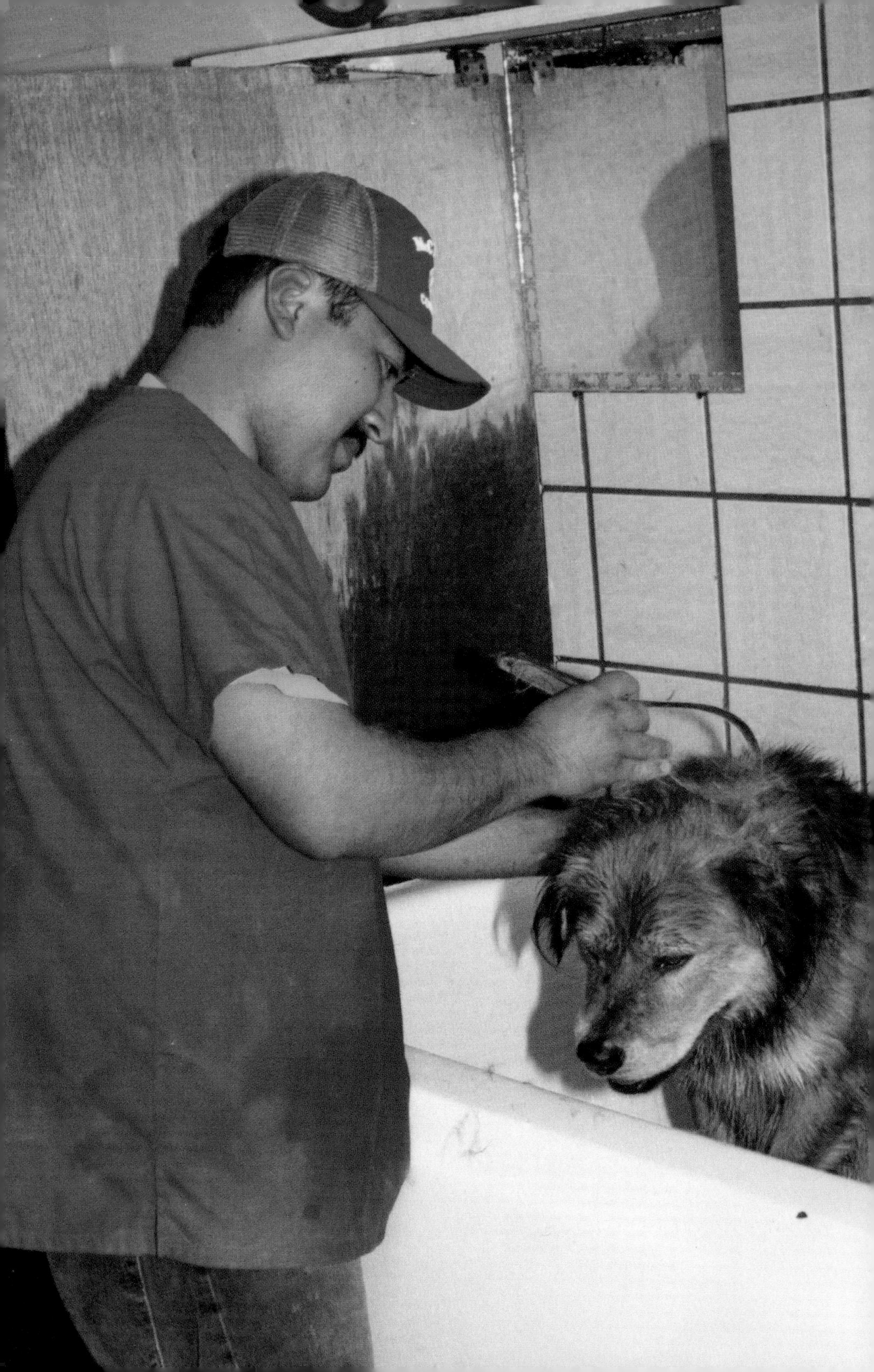

Chapter 1

Animal Caretaker

Animal caretakers care for animals. These animals may be pets, wild animals, laboratory animals, or even zoo animals. In Canada, animal caretakers are called animal care attendants.

Animal caretakers care for several types of animals. Many animal caretakers provide food, water, and shelter for wild or stray animals. Wild animals may live in zoos or wildlife refuges. Stray animals are animals that once were pets. These animals have been abandoned or lost by their owners. Some animal caretakers care for pets while their owners are traveling away from home.

Some animal caretakers care for pets while their owners are away.

What Animal Caretakers Do

Animal caretakers provide food and water for the animals under their care. They follow feeding schedules created by animal care experts. These schedules tell animal caretakers when to feed each animal. Animal caretakers measure food servings for each animal according to feeding instructions. They sometimes mix vitamins and medicines into animals' food.

Animal caretakers groom animals. They wash and brush animals. Caretakers clip animals' claws or nails and trim their fur.

Animal caretakers may exercise or train animals. Most animals must exercise to stay strong and healthy. Animals also may need physical activity to overcome and prevent illnesses. Animal caretakers sometimes exercise animals to train them. They train the animals to perform certain tasks or to behave in desired ways. This requires a great deal of patience.

Animal caretakers provide food and water for the animals under their care.

Animal caretakers observe animals for signs of illness or injury. They may provide basic treatment to sick or injured animals. Animal caretakers sometimes help veterinarians treat sick or injured animals. Animal caretakers often help control and calm animals so veterinarians can perform their work. Animal caretakers even may give vaccinations to animals. These medications are injected into

Animal caretakers brush animals.

animals to prevent them from getting certain diseases.

Animal caretakers often maintain animals' living quarters. They clean pens, stables, cages, and yards. Animal caretakers set up feeding, exercise, and heating equipment in animals' living quarters. They also may repair fences, cages, and pens.

Animal caretakers often keep records about the animals under their care. They record information such as animals' height, weight, diet, behavior, and medication.

Animal caretakers often work with other people. They talk to pet owners or other animal workers about animal behavior and diet. They give people advice about caring for and feeding animals.

Where Animal Caretakers Work

Animal caretakers can work in many settings. Some work at boarding kennels, grooming shops, or stables. They may care for animals in veterinary facilities, animal shelters, and research labs. Others may work with animals in zoos and aquariums. Aquariums exhibit many types of fish and other ocean life.

Some animal caretakers work only indoors. They may work in grooming shops. Others may work in animal shelters or research labs. Animal caretakers in these places may work with caged animals.

Other animal caretakers work outdoors. They work in all types of weather conditions. Animals in kennels and stables must be exercised even in cold and unpleasant weather. Animal caretakers who work at zoos and aquariums may care for animals that live outdoors.

Some animal caretakers work in grooming shops.

Chapter 2

Day-to-Day Activities

Animal caretakers' daily duties can vary. These duties depend on the type of animal care caretakers provide and their work settings.

Kennel Attendants

Animal caretakers who work in boarding kennels are called kennel attendants. Kennel attendants provide housing and care to pets when owners cannot. They give pets food, water, shelter, attention, and exercise.

Kennel attendants usually begin their work early in the morning. They clean each cage and dog run. These large indoor or outdoor pens

Animal caretakers' daily duties depend on the type of animal care they provide.

Kennel attendants watch animals for signs of illness or injury.

are designed to allow dogs to move freely. Kennel attendants change the paper used to line the cages. They clean cat litter boxes. Kennel attendants also may clean ferret or rabbit cages.

Kennel attendants give the animals attention. They hold and pet cats. They play with dogs and take them for walks.

Kennel attendants finish their morning routine by giving each animal clean water and fresh food. The routine is repeated in the afternoon or early evening.

Kennel attendants watch animals for signs of illness or injury. They watch for vomiting or changes in animals' waste. They report any concerns to a supervisor. These people direct animal caretakers' work. Kennel attendants sometimes give medicines to animals according to supervisors' directions.

Stable Workers

Animal caretakers who work with horses are called stable workers. Stable workers care for horses at riding stables, horse farms, and racetracks. They clean horses' stables. They give food and water to the horses. They also help train and exercise the horses.

Stable workers sometimes must move horses. They load horses into trailers and drive them to racetracks or horse shows.

Stable workers also perform other duties. They may clean and repair riding and racing equipment. Stable workers sometimes act as trail guides for people who ride horses.

Groomers

Animal caretakers who specialize in caring for pets' appearances are called groomers. Some groomers work in kennels or animal clinics. Other groomers run their own grooming businesses.

Groomers talk to pet owners at the beginning of each appointment. The groomer asks the owner how the owner would like the pet to look. The groomer and the owner also discuss the pet's behavior. Groomers must understand animal behavior. They also must be prepared for how pets may react while being groomed. Grooming procedures may frighten pets. Groomers may need to restrain pets to prevent them from biting or scratching.

Groomers then begin the grooming process. They brush tangles out of pets' fur. They then clip the fur. Groomers use combs, electric clippers, and grooming shears for this clipping. Groomers then cut pets' nails or claws and clean their ears. They then bathe the pets and blow-dry their fur. Groomers finish the grooming process with a final clipping and styling.

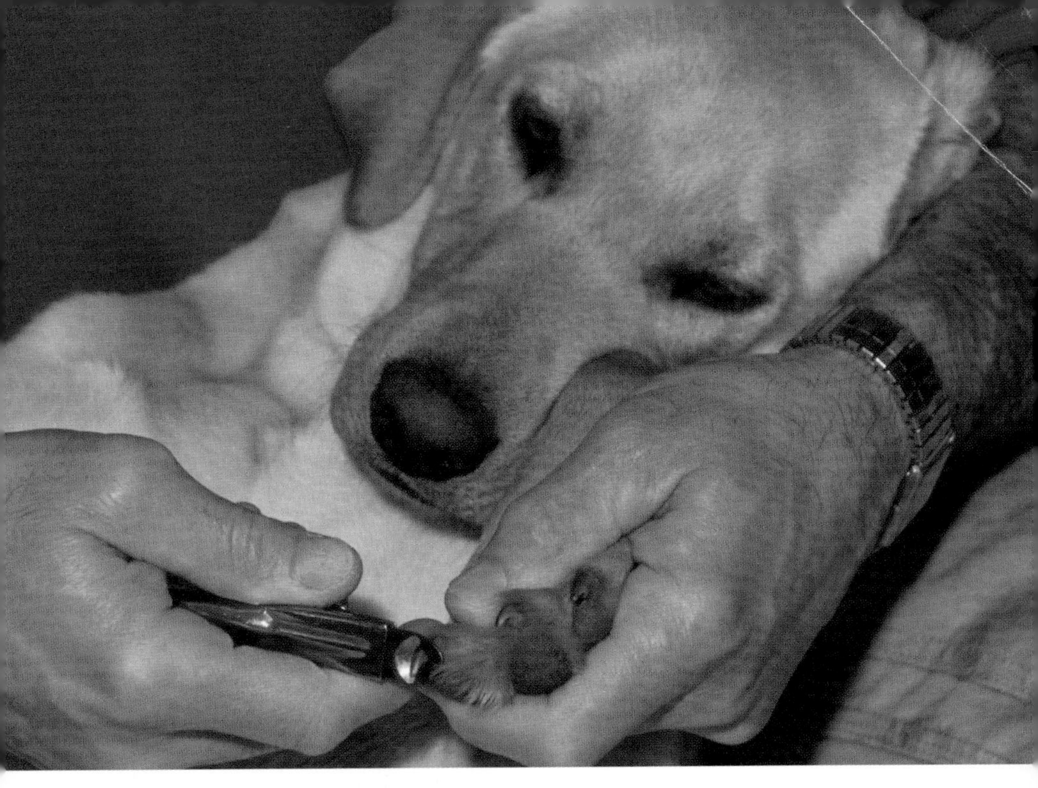

Groomers cut animals' nails or claws.

Veterinary Assistants

Animal caretakers in animal hospitals and clinics are called veterinary assistants. They prepare animals and equipment for medical procedures. They clean examination rooms and tables. They keep surgery areas and surgical instruments free of germs.

Veterinary assistants help veterinarians during routine procedures. They maintain control of the animals. They hold animals still for anesthesia. Anesthesia prevents animals

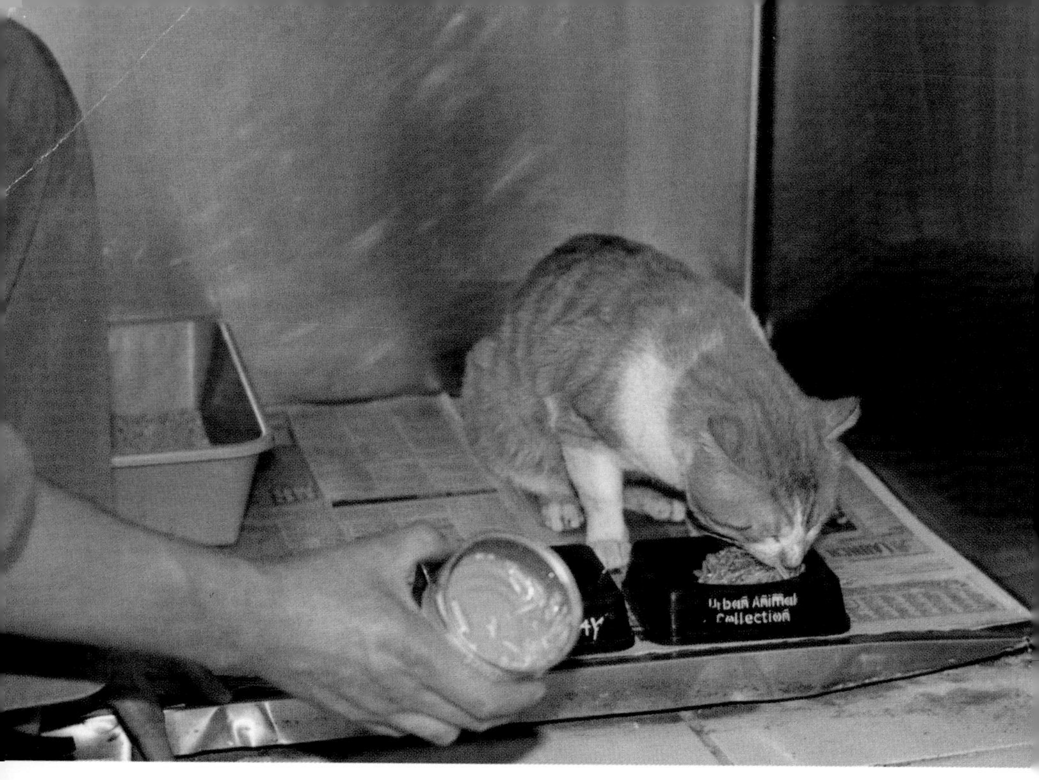

Animal caretakers in animal shelters provide basic care for animals.

from feeling pain during uncomfortable procedures and surgeries. Anesthesia sometimes can cause animals to have breathing problems or vomit. Veterinary assistants also clean work areas after the procedures are completed.

Veterinary assistants also observe animals after surgery. They check dressings. They keep these bandages secure and clean. They notify the veterinarian if any problems occur.

Veterinary assistants also may perform some basic procedures. They may clip animals' nails or claws. They may bathe animals in medicated baths to kill fleas. They may give medications to animals as directed by a veterinarian.

Animal Caretakers in Animal Shelters

Animal caretakers in animal shelters care for homeless animals. Animals in shelters may be strays or animals that are no longer wanted by their owners.

Animal caretakers at shelters have many duties. They maintain records of the animals that come in and out of the shelters. Many animals leave shelters when their owners claim them. Some leave when new owners adopt them. Others may be euthanized. Workers euthanize animals by injecting them with substances that stop the animals' breathing or heartbeat. These animals may be very sick or pose a danger to the public.

Animal caretakers in animal shelters begin each day by walking through the shelter. They check the condition of each animal. Caretakers remove all the animals from the area in order to clean the animals' living quarters. They then give food and water to the animals. Caretakers repeat this cleaning and feeding routine in the evening.

Animal caretakers in animal shelters often work with the public. Shelter animals sometimes are claimed by their owners. Animal caretakers then bring the animal to an area to meet its owner. Caretakers also decide who can adopt shelter animals that are not claimed. They determine if people can provide good homes for animals.

Animal Caretakers in Research Facilities

Animal caretakers in research facilities often work with animals used in medical research. They may care for rats, dogs, pigs, rabbits, or even monkeys.

Animal caretakers in research labs have many duties. They clean cages and living quarters each morning. They observe animals for signs of disease, injury, illness, or pain. They give food and water to the animals. Animals sometimes are allowed to eat only a specific type or amount of

Animal caretakers in shelters decide if people can adopt the animals in the shelter.

food. Animal caretakers feed animals according to specific instructions of supervisors. They also may give animals medicines according to these instructions.

Animal caretakers record information about laboratory animals. They record what and how much each animal eats. They record each animal's weight and details about any medicines they receive.

Zookeepers clean animals' living quarters.

Zookeepers and Aquarists

Animal caretakers who work in zoos are called zookeepers. Zookeepers care for a variety of wild animals. Zookeepers sometimes specialize in certain types of animals such as birds or reptiles. Zookeepers usually get to know the animals they care for very well.

Zookeepers perform many duties. They clean animals' living quarters. They prepare the animals' food. They note the animals' eating patterns and any changes in their behavior. Zookeepers observe the animals to make sure they are healthy. They watch for and record any signs of illness or injury.

Animal caretakers who work in aquariums are called aquarists. Aquarists feed and care for fish and other animals that live in the water. They observe the animals for signs of health problems. They clean aquariums and keep them in proper working order. They maintain water quality for the specific types of animals that live in each aquarium.

Zookeepers and aquarists often interact with the public. They provide information and answer questions about the animals in their care. They want people to enjoy watching and learning about animals. But they also want to protect the animals in their care. They warn people to act properly around the animals.

Chapter 3

The Right Candidate

People who want to be animal caretakers should enjoy working with animals. Animal caretakers must be concerned about each animal in their care.

Abilities

Animal caretakers must be logical. They use clear, concise thinking to learn the types of tasks involved in animal care. They also must understand animals' habits and physical needs. They observe animals for changes in appearance, appetite, or behavior. Changes may mean that something is wrong. Animal caretakers may need to determine if animals need medical care.

People who want to be animal caretakers should enjoy working with animals.

Animal caretakers must be in good physical condition to use tools such as high-pressure hoses.

Animal caretakers must be in good physical condition. They sometimes must handle frightened or upset animals. Animal caretakers also sometimes clean stables, kennels, and other living quarters. They use tools such as shovels, pitchforks, rakes, mops, and high-pressure hoses. They must be strong to use these tools. Animal caretakers also exercise animals. Some animal caretakers may walk several dogs each day.

Animal caretakers must be patient. Animals often do not understand people's actions. They may kick, bite, or scratch animal caretakers if they are frightened or injured. Animal caretakers must stay calm in such situations. They should remain patient when training animals. They often must repeat the same routines with an animal many times.

Basic Skills

Animal caretakers should have good motor skills. They may have to trim cat claws and dog nails without hurting the animals. They may fasten shoes on horses. Horses wear these curved pieces of metal to protect their hooves. Animal caretakers may give shots to animals. They may help veterinarians with more complicated medical procedures such as surgery.

Animal caretakers must be able to handle animals. They should be gentle so they do not frighten or hurt animals. Caretakers need to be strong and firm so they can control animals' movements when needed.

Skills

Workplace Skills

	Yes	No
Resources:		
Assign use of time	✓	☐
Assign use of money	☐	✓
Assign use of material and facility resources	✓	☐
Assign use of human resources	✓	☐
Interpersonal Skills:		
Take part as a member of a team	✓	☐
Teach others	✓	☐
Serve clients/customers	✓	☐
Show leadership	✓	☐
Work with others to arrive at a decision	✓	☐
Work with a variety of people	✓	☐
Information:		
Acquire and judge information	✓	☐
Understand and follow legal requirements	✓	☐
Organize and maintain information	✓	☐
Understand and communicate information	✓	☐
Use computers to process information	✓	☐
Systems:		
Identify, understand, and work with systems	✓	☐
Understand environmental, social, political, economic, or business systems	✓	☐
Oversee and correct system performance	☐	✓
Improve and create systems	☐	✓
Technology:		
Select technology	✓	☐
Apply technology to task	✓	☐
Maintain and troubleshoot technology	☐	✓

Foundation Skills

	Yes	No
Basic Skills:		
Read	✓	☐
Write	✓	☐
Do arithmetic and math	✓	☐
Speak and listen	✓	☐
Thinking Skills:		
Learn	✓	☐
Reason	✓	☐
Think creatively	✓	☐
Make decisions	✓	☐
Solve problems	✓	☐
Personal Qualities:		
Take individual responsibility	✓	☐
Have self-esteem and self-management	✓	☐
Be sociable	✓	☐
Be fair, honest, and sincere	✓	☐

Animal caretakers need basic math skills. They calculate and measure amounts of food. They measure medications based on animals' weight. Animal caretakers also maintain feeding and growth records for animals.

Animal caretakers also need basic language skills. They make written notes in their records about changes in animals' eating habits or behaviors. They discuss health concerns with veterinarians. Animal caretakers interact with the public in different work settings. They answer questions about animal behavior. They may give people suggestions about training animals. They advise people how to care for animals.

Work Styles

Animal caretakers must be willing to work in environments that sometimes are unpleasant. Animal caretakers clean up animal living quarters and animal waste. This waste may cause living quarters to smell bad. Animal caretakers may work outdoors in different weather conditions. They sometimes are exposed to diseased animals.

Animal caretaker workplaces sometimes can be noisy. Animals may make loud noises when they are in cages or around other animals. Animal caretakers must not be bothered by continuous loud noises.

Chapter 4

Preparing for the Career

Training for animal caretaker positions differs with each type of job. Most animal caretaker positions do not require formal training.

High School

People who want to work as animal caretakers usually do not need a high school diploma. But most employers prefer to hire animal caretakers with a high school diploma. Many high school courses help students prepare for animal care careers.

Science courses are useful to students interested in animal caretaker careers. Students

High school students interested in animal caretaker careers may volunteer at zoos.

Students who want to become animal caretakers may benefit from math courses.

study animals in zoology classes. In biology classes, students study many forms of living things. In chemistry classes, students learn about the features of many substances and how they interact with each other. This knowledge helps animal caretakers care for animals. Animal caretakers often give medications to animals. Caretakers must know how these medications work and interact with each other.

Students who want to become animal caretakers may benefit from math courses. Math skills help animal caretakers calculate animals' feeding portions based on written weight guidelines. These skills also help animal caretakers measure food and medications.

Students interested in animal caretaker careers may benefit from English and speech courses. These courses help students communicate clearly with others. Animal caretakers must be able to communicate with veterinarians, co-workers, and pet owners. Writing skills also help animal caretakers maintain written records about the animals in their care.

High school students also may benefit from holding part-time or summer jobs in animal care. Students may work at kennels or zoos. They also may acquire related work experience by volunteering at local animal care facilities. Students may earn little or no money from these positions. But they will gain valuable experience in animal care.

Some employers may prefer to hire animal caretakers who have experience working with animals.

Experience and Training

Many employers prefer to hire animal caretakers who have some experience working with animals. Some animal caretakers have experience working with farm animals. These workers are familiar with animal feeding schedules and maintenance of animal living quarters. Other animal caretakers have experience raising pets. They know that animals need affection in addition to basic care.

Most animal caretakers train on the job. But some training programs are available for certain animal caretaker jobs. Most pet groomers complete an informal apprenticeship under the guidance of experienced groomers. Beginning kennel attendants usually begin with simple tasks and move to more complicated tasks as they gain experience. Veterinary assistants are trained on the job under the guidance of veterinarians or veterinary technicians. Other types of animal caretakers also train on the job.

Chapter 5

The Market

Employment opportunities for animal caretakers are expected to grow faster than average. Many people keep animals as pets. People also enjoy observing animals in zoo and aquarium settings. These animals need daily care.

Salary and Job Outlook

In the United States, the average salary range for animal caretakers is about $8,840 to $25,000 per year. The average salary is about $14,300 per year. Average salaries for caretakers in animal control agencies or shelters are about $17,100 per year.

Employment opportunities for animal caretakers are expected to grow faster than average.

Stable workers can attend a training program to advance their careers.

In Canada, salaries for most animal care attendants are between $12,480 and $35,360 per year. The average salary for workers in this field is about $24,000.

Employment opportunities for animal caretakers in the United States and Canada are expected to be good. The popularity of pets and zoos continues to grow. Animal caretakers will be needed to care for these animals.

Advancement Opportunities

Animal caretakers have many opportunities to advance into related jobs. Advancement usually requires formal training or education.

Groomers can attend a formal six- to 16-week program at a technical school. They become certified by the National Dog Groomers Association of America after passing testing requirements. With this training, they may advance to senior groomer or management positions. They even may open their own grooming businesses.

Animal caretakers who work as kennel attendants may decide to operate their own kennels. They can take a three-part, home-study program offered by the American Boarding Kennels Association. They become Certified Kennel Operators after passing a test.

Animal caretakers who work as veterinary assistants can become veterinary technicians. These caretakers must complete a two-year veterinary program at a technical school or college.

Animal caretakers who work in research facilities may advance to positions in laboratory animal science. They can earn certification as a research assistant, mid-level technician, or senior-level technologist. They first must complete a university or vocational college program in laboratory animal science. They then apply for certification from the American Association for Laboratory Animal Science.

Zookeepers or aquarists may advance to supervisory jobs. For example, they can work as senior keepers, head keepers, or assistant curators. A bachelor's degree in biology or animal science usually is required for these positions. People usually earn bachelor's degrees in four years from a college or university.

Animal caretakers who work in animal shelters also may train for advancement. Humane organizations or the government offer training programs or workshops. Animal caretakers may advance to positions as adoption coordinators, assistant shelter managers, or shelter directors.

Related Careers

People who are interested in animals may find jobs in other career fields. People may work in

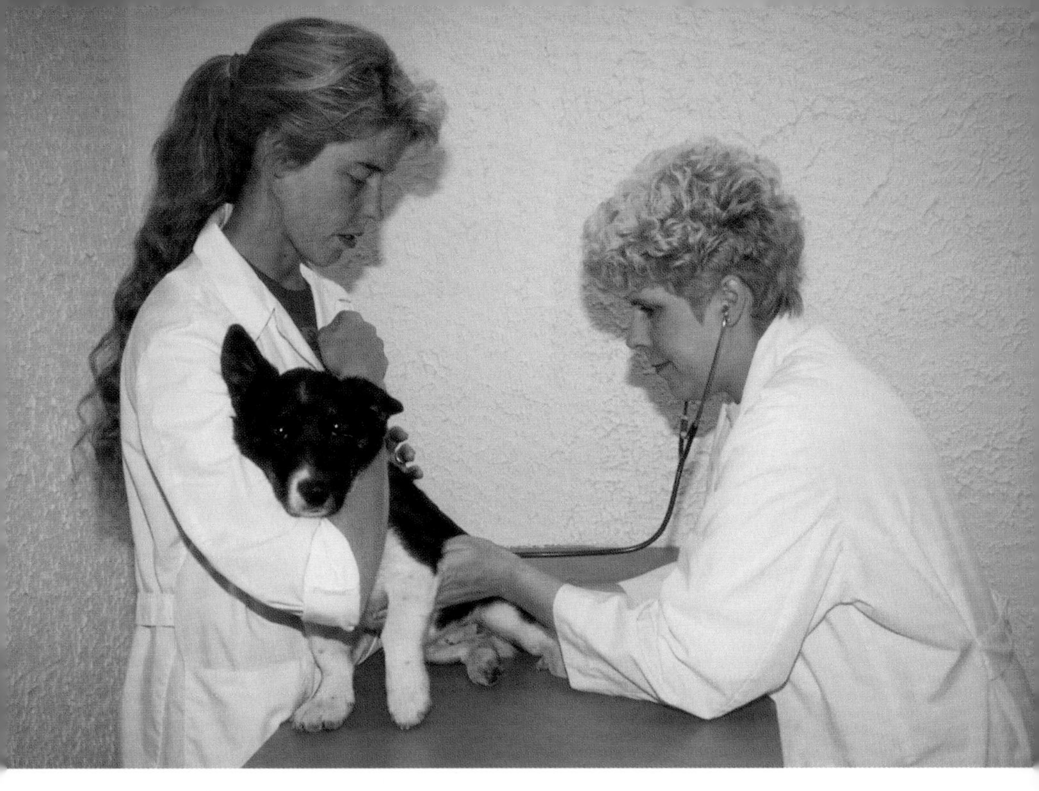

Some people who are interested in animals may become veterinarians.

pet stores, on ranches, or as animal breeders or trainers. Other people may become veterinarians. These people must earn Doctor of Veterinary Medicine (D.V.M.) degrees. This degree usually requires eight years of college education.

Animals play an important role in people's lives. Animal caretakers will continue to be needed to help these animals lead long, healthy lives.

Words to Know

anesthesia (an-iss-THEE-zhuh)—a gas or injection that prevents pain during treatments and operations

aquarist (uh-KWAIR-ist)—a person who works at an aquarium

boarding kennel (BORD-ing KEN-uhl)—a place where animals are looked after while their owners are away

euthanize (YOO-thuh-nize)—to painlessly put an animal to death by injecting it with a substance that stops its breathing or heartbeat

groom (GROOM)—to brush and clean an animal

vaccination (vak-suh-NAY-shun)—a shot of medicine that protects a person or animal from disease

veterinarian (vet-ur-uh-NER-ee-uhn)—a doctor who is trained to treat the illnesses and injuries of animals

To Learn More

Burgan, Michael. *Veterinarian.* Career Exploration. Mankato, Minn.: Capstone Books, 2000.

Cosgrove, Holli. *Career Discovery Encyclopedia.* Vol. 1. Chicago: Ferguson Publishing, 2000.

Hurwitz, Jane. *Choosing a Career in Animal Care.* World of Work. New York: Rosen Publishing Group, 1997.

Lee, Mary Price and Richard S. Lee. *Opportunities in Animal and Pet Care Careers.* VGM Opportunities. Lincolnwood, Ill.: VGM Career Horizons, 1994.

Miller, Louise. *Animals.* VGM Career Portraits. Lincolnwood, Ill.: VGM Career Horizons, 1995.

Useful Addresses

American Boarding Kennels Association
4575 Galley Road
Suite 400A
Colorado Springs, CO 80915

Canadian Veterinary Medical Association
339 Booth Street
Ottawa, ON K1R 7K1
Canada

Humane Society of the United States
2100 L Street NW
Washington, DC 20037

National Animal Control Association
P.O. Box 480851
Kansas City, MO 64148

Internet Sites

American Veterinary Assistants' Association
http://www.avaa.bigstep.com

Job Futures—Other Occupations in Personal Services
http://www.jobfutures.ca/jobfutures/noc/648.html

NetVet Veterinary Resources The Electronic Zoo
http://netvet.wustl.edu

Occupational Outlook Handbook—Animal Caretakers and Service Workers
http://stats.bls.gov/oco/ocos168.htm

Index

advancement, 41–42
anesthesia, 19, 20
animal clinics, 18
apprenticeship, 37
aquarist, 24–25, 42
aquarium, 11, 13, 25

boarding kennels, 11, 15

claws, 8, 18, 21, 29

education, 33–35, 41, 43

groomer, 18, 37, 41

illness, 8, 9, 17, 22, 25
injury, 9, 17, 22, 25

medications, 8, 9, 11, 17, 21, 23, 31, 34, 35

nails, 8, 18, 21, 29

pets, 7, 15, 18, 37, 39, 40

research facility, 22–23, 41

salary, 39–40
skills, 29, 31, 35
stables, 10, 11, 13, 17, 28
stray animals, 7, 21

training, 33, 36–37, 41

vaccinations, 9
veterinarian, 9, 19, 20, 21, 29, 31, 35, 37, 43
veterinary assistant, 19–21, 37, 41

wild animals, 7, 24

zookeeper, 24–25, 42
zoos, 7, 11, 13, 24, 35, 40